CASUALTY INVESTIGATION CODE

CODE OF THE INTERNATIONAL STANDARDS AND
RECOMMENDED PRACTICES FOR A SAFETY INVESTIGATION
INTO A MARINE CASUALTY OR MARINE INCIDENT

2008 Edition

London, 2008

First published in 2008
by the INTERNATIONAL MARITIME ORGANIZATION
4 Albert Embankment, London SE1 7SR
www.imo.org

Printed in the United Kingdom by CPI Books Limited, Reading RG1 8EX

ISBN: 978-92-801-1498-0

IMO PUBLICATION

Sales number: I128E

Copyright © International Maritime Organization 2008

All rights reserved.
No part of this publication may be reproduced,
stored in a retrieval system or transmitted in any form
or by any means without prior permission in writing
from the International Maritime Organization.

This publication has been prepared from official documents of IMO, and every effort has been made to eliminate errors and reproduce the original text(s) faithfully. Readers should be aware that, in case of inconsistency, the official IMO text will prevail.

Foreword

1 This Code incorporates and builds on the best practices in marine casualty and marine incident investigation that were established by the Code for the Investigation of Marine Casualties and Incidents, adopted in November 1997 by the International Maritime Organization (the Organization), by resolution A.849(20). The Code for the Investigation of Marine Casualties and Incidents sought to promote co-operation and a common approach to marine casualty and marine incident investigations between States.

Background

2 The Organization has encouraged co-operation and recognition of mutual interest through a number of resolutions. The first was resolution A.173(ES.IV) (Participation in Official Inquiries into Maritime Casualties) adopted in November 1968. Other resolutions followed including: resolution A.322(IX) (The Conduct of Investigations into Casualties) adopted in November 1975; resolution A.440(XI) (Exchange of Information for Investigations into Marine Casualties) and resolution A.442(XI) (Personnel and Material Resource Needs of Administrations for the Investigation of Casualties and the Contravention of Conventions), both adopted in November 1979; and resolution A.637(16) (Co-operation in Maritime Casualty Investigations) adopted in 1989.

3 These individual resolutions were amalgamated and expanded by the Organization with the adoption of the Code for the Investigation of Marine Casualties and Incidents. Resolution A.884(21) (Amendments to the Code for the Investigation of Marine Casualties and Incidents resolution A.849(20)), adopted in November 1999, enhanced the Code by providing guidelines for the investigation of human factors.

4 The International Convention for the Safety of Life at Sea (SOLAS), 1948, included a provision requiring flag State Administrations to conduct investigations into any casualty suffered by a ship of its flag if an investigation may assist in identifying regulatory issues as a contributing factor. This provision was retained in the 1960 and 1974 SOLAS Conventions. It was also included in the International Convention on Load Lines, 1966. Further, flag States are required to inquire into certain marine casualties and marine incidents occurring on the high seas.*

---

* Reference is made to the United Nations Convention on the Law of Sea (UNCLOS), article 94(7) or requirements of international and customary laws.

Foreword

5 The sovereignty of a coastal State extends beyond its land and inland waters to the extent of its territorial sea.* This jurisdiction gives the coastal State an inherent right to investigate marine casualties and marine incidents connected with its territory. Most national Administrations have legal provisions to cover the investigation of a shipping incident within its inland waters and territorial sea, regardless of the flag.

Treatment of seafarers

6 Most recently, the International Labour Organization's Maritime Labour Convention, 2006 (which has not yet come into force), provides a provision for the investigation of some serious marine casualties as well as setting out working conditions for seafarers. Recognizing the need for special protection for seafarers during an investigation, the Organization adopted, in December 2005, the Guidelines on fair treatment of seafarers in the event of a maritime accident through resolution A.987(24). The Guidelines were promulgated by IMO and ILO on 1 July 2006.

Adoption of the Code

7 Since the adoption of the first SOLAS Convention, there have been extensive changes in the structure of the international maritime industry and changes in international law. These changes have potentially increased the number of States with an interest in the process and outcomes of marine safety investigations, in the event of a marine casualty or marine incident, increasing the potential for jurisdictional and other procedural differences between affected States.

8 This Code, while it specifies some mandatory requirements, recognizes the variations in international and national laws in relation to the investigation of marine casualties and marine incidents. The Code is designed to facilitate objective marine safety investigations for the benefit of flag States, coastal States, the Organization and the shipping industry in general.

---

* Reference is made to the United Nations Convention on the Law of the Sea (UNCLOS), article 2 or requirements of international and customary laws.

Contents

Page

PART I – GENERAL PROVISIONS
Chapter 1 – Purpose 1
Chapter 2 – Definitions 2
Chapter 3 – Application of chapters in Parts II and III 6

PART II – MANDATORY STANDARDS
Chapter 4 – Marine safety investigation Authority 7
Chapter 5 – Notification 7
Chapter 6 – Requirement to investigate very serious marine casualties 8
Chapter 7 – Flag State's agreement with another substantially interested State to conduct a marine safety investigation 8
Chapter 8 – Powers of an investigation 9
Chapter 9 – Parallel investigations 9
Chapter 10 – Co-operation 10
Chapter 11 – Investigation not to be subject to external direction 10
Chapter 12 – Obtaining evidence from seafarers 10
Chapter 13 – Draft marine safety investigation reports 11
Chapter 14 – Marine safety investigation reports 12

PART III – RECOMMENDED PRACTICES
Chapter 15 – Administrative responsibilities 13
Chapter 16 – Principles of investigation 13
Chapter 17 – Investigation of marine casualties (other than very serious marine casualties) and marine incidents 15

Contents

	Page
Chapter 18 – Factors that should be taken into account when seeking agreement under chapter 7 of Part II	15
Chapter 19 – Acts of unlawful interference	16
Chapter 20 – Notification to parties involved and commencement of an investigation	16
Chapter 21 – Co-ordinating an investigation	17
Chapter 22 – Collection of evidence	18
Chapter 23 – Confidentiality of information	19
Chapter 24 – Protection for witnesses and involved parties	20
Chapter 25 – Draft and final report	21
Chapter 26 – Re-opening an investigation	22

Resolution MSC.255(84)
adopted on 16 May 2008 23

Code of the International Standards and Recommended Practices for a Safety Investigation into a Marine Casualty or Marine Incident*

Part I
GENERAL PROVISIONS

Chapter 1
Purpose

1.1 The objective of this Code is to provide a common approach for States to adopt in the conduct of marine safety investigations into marine casualties and marine incidents. Marine safety investigations do not seek to apportion blame or determine liability. Instead a marine safety investigation, as defined in this Code, is an investigation conducted with the objective of preventing marine casualties and marine incidents in the future. The Code envisages that this aim will be achieved through States:

.1 applying consistent methodology and approach, to enable and encourage a broad ranging investigation, where necessary, in the interests of uncovering the causal factors and other safety risks; and

.2 providing reports to the Organization to enable a wide dissemination of information to assist the international marine industry to address safety issues.

1.2 A marine safety investigation should be separate from, and independent of, any other form of investigation. However, it is not the purpose of this Code to preclude any other form of investigation, including investigations for action in civil, criminal and administrative proceedings. Further, it is not the intent of the Code for a State or States conducting a

---

* The Code of the International Standards and Recommended Practices for a Safety Investigation into a Marine Casualty or Marine Incident (Casualty Investigation Code) comprises the annex to resolution MSC.255(84), the text of which is reproduced at the end of the present publication.

marine safety investigation to refrain from fully reporting on the causal factors of a marine casualty or marine incident because blame or liability, may be inferred from the findings.

1.3 This Code recognizes that under the Organization's instruments, each flag State has a duty to conduct an investigation into any casualty occurring to any ship flying its flag, when it judges that such an investigation may assist in determining what changes in the present regulations may be desirable, or if such a casualty has produced a major deleterious effect upon the environment. The Code also takes into account that a flag State shall* cause an inquiry to be held, by or before a suitably qualified person or persons into certain marine casualties or marine incidents of navigation on the high seas. However, the Code also recognizes that where a marine casualty or marine incident occurs within the territory, including the territorial sea, of a State, that State has a right[†] to investigate the cause of any such marine casualty or marine incident which might pose a risk to life or to the environment, involve the coastal State's search and rescue authorities, or otherwise affect the coastal State.

Chapter 2
Definitions

When the following terms are used in the mandatory standards and recommended practices for marine safety investigations they have the following meaning.

2.1 An *agent* means any person, natural or legal, engaged on behalf of the owner, charterer or operator of a ship, or the owner of the cargo, in providing shipping services, including managing arrangements for the ship being the subject of a marine safety investigation.

2.2 A *causal factor* means actions, omissions, events or conditions, without which:

    .1 the marine casualty or marine incident would not have occurred; or

---

* Reference is made to the United Nations Convention on the Law of Sea (UNCLOS), article 94 or requirements of international and customary laws.
† Reference is made to the United Nations Convention on the Law of Sea (UNCLOS), article 2 or requirements of international and customary laws.

.2 adverse consequences associated with the marine casualty or marine incident would probably not have occurred or have been as serious;

.3 another action, omission, event or condition, associated with an outcome in .1 or .2, would probably not have occurred.

2.3 A *coastal State* means a State in whose territory, including its territorial sea, a marine casualty or marine incident occurs.

2.4 *Exclusive economic zone* means the exclusive economic zone as defined by article 55 of the United Nations Convention on the Law of the Sea.

2.5 *Flag State* means a State whose flag a ship is entitled to fly.

2.6 *High seas* means the high seas as defined in article 86 of the United Nations Convention on the Law of the Sea.

2.7 *Interested party* means an organization, or individual, who, as determined by the marine safety investigating State(s), has significant interests, rights or legitimate expectations with respect to the outcome of a marine safety investigation.

2.8 *International Safety Management (ISM) Code* means the International Management Code for the Safe Operation of Ships and for Pollution Prevention as adopted by the Organization by resolution A.741(18), as amended.

2.9 A *marine casualty* means an event, or a sequence of events, that has resulted in any of the following which has occurred directly in connection with the operations of a ship:

.1 the death of, or serious injury to, a person;

.2 the loss of a person from a ship;

.3 the loss, presumed loss or abandonment of a ship;

.4 material damage to a ship;

.5 the stranding or disabling of a ship, or the involvement of a ship in a collision;

.6 material damage to marine infrastructure external to a ship, that could seriously endanger the safety of the ship, another ship or an individual; or

.7 severe damage to the environment, or the potential for severe damage to the environment, brought about by the damage of a ship or ships.

However, a marine casualty does not include a deliberate act or omission, with the intention to cause harm to the safety of a ship, an individual or the environment.

2.10 A *marine incident* means an event, or sequence of events, other than a marine casualty, which has occurred directly in connection with the operations of a ship that endangered, or, if not corrected, would endanger the safety of the ship, its occupants or any other person or the environment.

However, a marine incident does not include a deliberate act or omission, with the intention to cause harm to the safety of a ship, an individual or the environment.

2.11 A *marine safety investigation* means an investigation or inquiry (however referred to by a State), into a marine casualty or marine incident, conducted with the objective of preventing marine casualties and marine incidents in the future. The investigation includes the collection of, and analysis of, evidence, the identification of causal factors and the making of safety recommendations as necessary.

2.12 A *marine safety investigation report* means a report that contains:

.1 a summary outlining the basic facts of the marine casualty or marine incident and stating whether any deaths, injuries or pollution occurred as a result;

.2 the identity of the flag State, owners, operators, the company as identified in the safety management certificate, and the classification society (subject to any national laws concerning privacy);

.3 where relevant the details of the dimensions and engines of any ship involved, together with a description of the crew, work routine and other matters, such as time served on the ship;

.4 a narrative detailing the circumstances of the marine casualty or marine incident;

.5 analysis and comment on the causal factors including any mechanical, human and organizational factors;

.6 a discussion of the marine safety investigation's findings, including the identification of safety issues, and the marine safety investigation's conclusions; and

.7 where appropriate, recommendations with a view to preventing future marine casualties and marine incidents.

2.13 *Marine safety investigation Authority* means an Authority in a State, responsible for conducting investigations in accordance with this Code.

Chapter 2

2.14 *Marine safety investigating State(s)* means the flag State or, where relevant, the State or States that take the responsibility for the conduct of the marine safety investigation as mutually agreed in accordance with this Code.

2.15 A *marine safety record* means the following types of records collected for a marine safety investigation:

.1 all statements taken for the purpose of a marine safety investigation;

.2 all communications between persons pertaining to the operation of the ship;

.3 all medical or private information regarding persons involved in the marine casualty or marine incident;

.4 all records of the analysis of information or evidential material acquired in the course of a marine safety investigation; and

.5 information from the voyage data recorder.

2.16 A *material damage* in relation to a marine casualty means:

.1 damage that:

.1.1 significantly affects the structural integrity, performance or operational characteristics of marine infrastructure or of a ship; and

.1.2 requires major repair or replacement of a major component or components; or

.2 destruction of the marine infrastructure or ship.

2.17 A *seafarer* means any person who is employed or engaged or works in any capacity on board a ship.

2.18 A *serious injury* means an injury which is sustained by a person, resulting in incapacitation where the person is unable to function normally for more than 72 hours, commencing within seven days from the date when the injury was suffered.

2.19 A *severe damage to the environment* means damage to the environment which, as evaluated by the State(s) affected, or the flag State, as appropriate, produces a major deleterious effect upon the environment.

2.20 *Substantially interested State* means a State:

.1 which is the flag State of a ship involved in a marine casualty or marine incident; or

5

.2 which is the coastal State involved in a marine casualty or marine incident; or

.3 whose environment was severely or significantly damaged by a marine casualty (including the environment of its waters and territories recognized under international law); or

.4 where the consequences of a marine casualty or marine incident caused, or threatened, serious harm to that State or to artificial islands, installations, or structures over which it is entitled to exercise jurisdiction; or

.5 where, as a result of a marine casualty, nationals of that State lost their lives or received serious injuries; or

.6 that has important information at its disposal that the marine safety investigating State(s) consider useful to the investigation; or

.7 that for some other reason establishes an interest that is considered significant by the marine safety investigating State(s).

2.21 *Territorial sea* means territorial sea as defined by Section 2 of Part II of the United Nations Convention on the Law of the Sea.

2.22 A *very serious marine casualty* means a marine casualty involving the total loss of the ship or a death or severe damage to the environment.

Chapter 3
Application of chapters in Parts II and III

3.1 Part II of this Code contains mandatory standards for marine safety investigations. Some clauses apply only in relation to certain categories of marine casualties and are mandatory only for marine safety investigations into those marine casualties.

3.2 Clauses in Part III of this Code may refer to clauses in this Part that apply only to certain marine casualties. The clauses in Part III may recommend that such clauses be applied in marine safety investigations into other marine casualties or marine incidents.

Part II
MANDATORY STANDARDS

Chapter 4
Marine safety investigation Authority

4.1 The Government of each State shall provide the Organization with detailed contact information of the marine safety investigation Authority (ies) carrying out marine safety investigations within their State.

Chapter 5
Notification

5.1 When a marine casualty occurs on the high seas or in an exclusive economic zone, the flag State of a ship, or ships, involved, shall notify other substantially interested States as soon as is reasonably practicable.

5.2 When a marine casualty occurs within the territory, including the territorial sea, of a coastal State, the flag State, and the coastal State, shall notify each other and between them notify other substantially interested States as soon as is reasonably practicable.

5.3 Notification shall not be delayed due to the lack of complete information.

5.4 **Format and content:** The notification shall contain as much of the following information as is readily available:

.1 the name of the ship and its flag State;
.2 the IMO ship identification number;
.3 the nature of the marine casualty;
.4 the location of the marine casualty;
.5 time and date of the marine casualty;
.6 the number of any seriously injured or killed persons;
.7 consequences of the marine casualty to individuals, property and the environment; and
.8 the identification of any other ship involved.

Chapter 6
Requirement to investigate very serious marine casualties

6.1   A marine safety investigation shall be conducted into every very serious marine casualty.

6.2   Subject to any agreement in accordance with chapter 7, the flag State of a ship involved in a very serious marine casualty is responsible for ensuring that a marine safety investigation is conducted and completed in accordance with this Code.

Chapter 7
Flag State's agreement with another substantially interested State to conduct a marine safety investigation

7.1   Without limiting the rights of States to conduct their own separate marine safety investigation, where a marine casualty occurs within the territory, including territorial sea, of a State, the flag State(s) involved in the marine casualty and the coastal State shall consult to seek agreement on which State or States will be the marine safety investigating State(s) in accordance with a requirement, or a recommendation acted upon, to investigate under this Code.

7.2   Without limiting the rights of States to conduct their own separate marine safety investigation, if a marine casualty occurs on the high seas or in the exclusive economic zone of a State, and involves more than one flag State, then the States shall consult to seek agreement on which State or States will be the marine safety investigating State(s) in accordance with a requirement, or a recommendation acted upon, to investigate under this Code.

7.3   For a marine casualty referred to in paragraph 7.1 or 7.2, agreement may be reached by the relevant States with another substantially interested State for that State or States to be the marine safety investigating State(s).

7.4 Prior to reaching an agreement, or if an agreement is not reached, in accordance with paragraph 7.1, 7.2 or 7.3, then the existing obligations and rights of States under this Code, and under other international laws, to conduct a marine safety investigation, remain with the respective parties to conduct their own investigation.

7.5 By fully participating in a marine safety investigation conducted by another substantially interested State, the flag State shall be considered to fulfil its obligations under this Code, SOLAS regulation I/21 and article 94, section 7 of the United Nations Convention on the Law of the Sea.

Chapter 8
Powers of an investigation

8.1 All States shall ensure that their national laws provide investigator(s) carrying out a marine safety investigation with the ability to board a ship, interview the master and crew and any other person involved, and acquire evidential material for the purposes of a marine safety investigation.

Chapter 9
Parallel investigations

9.1 Where the marine safety investigating State(s) is conducting a marine safety investigation under this Code, nothing prejudices the right of another substantially interested State to conduct its own separate marine safety investigation.

9.2 While recognizing that the marine safety investigating State(s) shall be able to fulfil obligations under this Code, the marine safety investigating State(s) and any other substantially interested State conducting a marine safety investigation shall seek to co-ordinate the timing of their investigations, to avoid conflicting demands upon witnesses and access to evidence, where possible.

Chapter 10
Co-operation

10.1 All substantially interested States shall co-operate with the marine safety investigating State(s) to the extent practicable. The marine safety investigating State(s) shall provide for the participation of the substantially interested States to the extent practicable.*

Chapter 11
Investigation not to be subject to external direction

11.1 Marine safety investigating State(s) shall ensure that investigator(s) carrying out a marine safety investigation are impartial and objective. The marine safety investigation shall be able to report on the results of a marine safety investigation without direction or interference from any persons or organizations that may be affected by its outcome.

Chapter 12
Obtaining evidence from seafarers

12.1 Where a marine safety investigation requires a seafarer to provide evidence to it, the evidence shall be taken at the earliest practical opportunity. The seafarer shall be allowed to return to his/her ship, or be repatriated at the earliest possible opportunity. The seafarer's human rights shall, at all times, be upheld.

12.2 All seafarers from whom evidence is sought shall be informed of the nature and basis of the marine safety investigation. Further, a seafarer from whom evidence is sought shall be informed, and allowed access to legal advice, regarding:

    .1 any potential risk that he/she may incriminate himself/herself in any proceedings subsequent to the marine safety investigation;

    .2 any right not to self-incriminate or to remain silent;

---

* The reference to "extent practicable" may be taken to mean, as an example, that co-operation or participation is limited because national laws make it impracticable to fully co-operate or participate.

.3 any protections afforded to the seafarer to prevent the evidence being used against him/her if he/she provides the evidence to the marine safety investigation.

Chapter 13
Draft marine safety investigation reports

13.1 Subject to paragraphs 13.2 and 13.3, where it is requested, the marine safety investigating State(s) shall send a copy of a draft report to a substantially interested State to allow the substantially interested State to comment on the draft report.

13.2 The marine safety investigating State(s) is only bound to comply with paragraph 13.1 where the substantially interested State receiving the report guarantees not to circulate, nor cause to circulate, publish or give access to the draft report, or any part thereof, without the express consent of the marine safety investigating State(s) or unless such reports or documents have already been published by the marine safety investigating State(s).

13.3 The marine safety investigating State(s) is not bound to comply with paragraph 13.1 if:

.1 the marine safety investigating State(s) requests the substantially interested State receiving the report to affirm that evidence included in the draft report will not be admitted in civil or criminal proceedings against a person who gave the evidence; and

.2 the substantially interested State refuses to provide such an affirmation.

13.4 The marine safety investigating State(s) shall invite the substantially interested States to submit their comments on the draft report within 30 days or some other mutually agreed period. The marine safety investigating State(s) shall consider the comments before preparing the final report and where the acceptance or rejection of the comments will have a direct impact on the interests of the State that submitted them, the marine safety investigating State(s) shall notify the substantially interested State of the manner in which the comments were addressed. If the marine safety investigating State(s) receives no comments after 30 days or the mutually agreed period has expired, then it may proceed to finalize the report.

13.5 The marine safety investigating State(s) shall seek to fully verify the accuracy and completeness of the draft report by the most practical means.

Chapter 14
Marine safety investigation reports

14.1 The marine safety investigating State(s) shall submit the final version of a marine safety investigation report to the Organization for every marine safety investigation conducted into a very serious marine casualty.

14.2 Where a marine safety investigation is conducted into a marine casualty or marine incident, other than a very serious marine casualty, and a marine safety investigation report is produced which contains information which may prevent or lessen the seriousness of marine casualties or marine incidents in the future, the final version shall be submitted to the Organization.

14.3 The marine safety investigation report referred in paragraphs 14.1 and 14.2 shall utilize all the information obtained during a marine safety investigation, taking into account its scope, required to ensure that all the relevant safety issues are included and understood so that safety action can be taken as necessary.

14.4 The final marine safety investigation report shall be made available to the public and the shipping industry by the marine safety investigating State(s), or the marine safety investigating State(s) shall undertake to assist the public and the shipping industry with details, necessary to access the report, where it is published by another State or the Organization.

Part III
RECOMMENDED PRACTICES

Chapter 15
Administrative responsibilities

15.1 States should ensure that marine safety investigating Authorities have available to them sufficient material and financial resources and suitably qualified personnel to enable them to facilitate the State's obligations to undertake marine safety investigations into marine casualties and marine incidents under this Code.

15.2 Any investigator forming part of a marine safety investigation should be appointed on the basis of the skills outlined in resolution A.996(25) for investigators.

15.3 However, paragraph 15.2 does not preclude the appropriate appointment of investigators with necessary specialist skills to form part of a marine safety investigation on a temporary basis, neither does it preclude the use of consultants to provide expert advice on any aspect of a marine safety investigation.

15.4 Any person who is an investigator, in a marine safety investigation, or assisting a marine safety investigation, should be bound to operate in accordance with this Code.

Chapter 16
Principles of investigation

16.1 **Independence:** A marine safety investigation should be unbiased to ensure the free flow of information to it.

16.1.1 In order to achieve the outcome in paragraph 16.1, the investigator(s) carrying out a marine safety investigation should have functional independence from:

   .1   the parties involved in the marine casualty or marine incident;
   .2   anyone who may make a decision to take administrative or disciplinary action against an individual or organization involved in a marine casualty or marine incident; and
   .3   judicial proceedings.

16.1.2 The investigator(s) carrying out a marine safety investigation should be free of interference from the parties in .1, .2 and .3 of paragraph 16.1.1 with respect to:

.1 the gathering of all available information relevant to the marine casualty or marine incident, including voyage data recordings and vessel traffic services recordings;

.2 analysis of evidence and the determination of causal factors;

.3 drawing conclusions relevant to the causal factors;

.4 distributing a draft report for comment and preparation of the final report; and

.5 if appropriate, the making of safety recommendations.

16.2 Safety focused: It is not the objective of a marine safety investigation to determine liability, or apportion blame. However, the investigator(s) carrying out a marine safety investigation should not refrain from fully reporting on the causal factors because fault or liability may be inferred from the findings.

16.3 Co-operation: Where it is practicable and consistent with the requirements and recommendations of this Code, in particular chapter 10 on co-operation, the marine safety investigating State(s) should seek to facilitate maximum co-operation between substantially interested States and other persons or organizations conducting an investigation into a marine casualty or marine incident.

16.4 Priority: A marine safety investigation should, as far as possible, be afforded the same priority as any other investigation, including investigations by a State for criminal purposes being conducted into the marine casualty or marine incident.

16.4.1 In accordance with paragraph 16.4 investigator(s) carrying out a marine safety investigation should not be prevented from having access to evidence in circumstances where another person or organization is carrying out a separate investigation into a marine casualty or marine incident.

16.4.2 The evidence for which ready access should be provided should include:

.1 survey and other records held by the flag State, the owners, and classification societies;

.2 all recorded data, including voyage data recorders; and

.3 evidence that may be provided by government surveyors, coastguard officers, vessel traffic service operators, pilots or other marine personnel.

16.5 Scope of a marine safety investigation: Proper identification of causal factors requires timely and methodical investigation, going far beyond the immediate evidence and looking for underlying conditions, which may be remote from the site of the marine casualty or marine incident, and which may cause other future marine casualties and marine incidents. Marine safety investigations should, therefore, be seen as a means of identifying not only immediate causal factors but also failures that may be present in the whole chain of responsibility.

Chapter 17
Investigation of marine casualties (other than very serious casualties) and marine incidents

17.1 A marine safety investigation should be conducted into marine casualties (other than very serious marine casualties – which are addressed in chapter 6 of this Code) and marine incidents, by the flag State of a ship involved, if it is considered likely that a marine safety investigation will provide information that can be used to prevent marine casualties and marine incidents in the future.

17.2 Chapter 7 contains the mandatory requirements for determining which is the marine safety investigating State(s) for a marine casualty. Where the occurrence being investigated in accordance with this chapter is a marine incident, chapter 7 should be followed as a recommended practice as if it referred to marine incidents.

Chapter 18
Factors that should be taken into account when seeking agreement under chapter 7 of Part II

18.1 When the flag State(s), a coastal State (if involved) or other substantially interested States are seeking to reach agreement, in accordance with chapter 7 of Part II on which State or State(s) will be the marine safety investigating State(s) under this Code, the following factors should be taken into account:

.1 whether the marine casualty or marine incident occurred in the territory, including territorial sea, of a State;

15

.2 whether the ship or ships involved in a marine casualty or marine incident occurring on the high seas, or in the exclusive economic zone, subsequently sail into the territorial sea of a State;

.3 the resources and commitment required of the flag State and other substantially interested States;

.4 the potential scope of the marine safety investigation and the ability of the flag State or another substantially interested State to accommodate that scope;

.5 the need of the investigator(s) carrying out a marine safety investigation to access evidence and consideration of the State or States best placed to facilitate that access to evidence;

.6 any perceived or actual adverse effects of the marine casualty or marine incident on other States;

.7 the nationality of the crew, passengers and other persons affected by the marine casualty or marine incident.

Chapter 19
Acts of unlawful interference

19.1 If in the course of a marine safety investigation it becomes known or is suspected that an offence is committed under article 3, 3*bis*, 3*ter* or 3*quater* of the Convention for the Suppression of Unlawful Acts Against the Safety of Maritime Navigation, 1988, the marine safety investigation Authority should immediately seek to ensure that the maritime security Authorities of the State(s) concerned are informed.

Chapter 20
Notification to parties involved and commencement of an investigation

20.1 When a marine safety investigation is commenced under this Code, the master, the owner and agent of a ship involved in the marine casualty or marine incident being investigated, should be informed as soon as practicable of:

.1 the marine casualty or marine incident under investigation;

.2 the time and place at which the marine safety investigation will commence;

.3 the name and contact details of the marine safety investigation Authority(ies);

.4 the relevant details of the legislation under which the marine safety investigation is being conducted;

.5 the rights and obligations of the parties subject to the marine safety investigation; and

.6 the rights and obligations of the State or States conducting the marine safety investigation.

20.2 Each State should develop a standard document detailing the information in paragraph 20.1 that can be transmitted electronically to the master, the agent and the owner of the ship.

20.3 Recognizing that any ship involved in a marine casualty or marine incident may continue in service, and that a ship should not be delayed more than is absolutely necessary, the marine safety investigating State(s) conducting the marine safety investigation should start the marine safety investigation as soon as is reasonably practicable, without delaying the ship unnecessarily.

Chapter 21
Co-ordinating an investigation

21.1 The recommendations in this chapter should be applied in accordance with the principles in chapters 10 and 11 of this Code.

21.2 The marine safety investigating State(s) should ensure that there is an appropriate framework within the State for:

.1 the designation of investigators to the marine safety investigation including an investigator to lead the marine safety investigation;

.2 the provision of a reasonable level of support to members of the marine safety investigation;

.3 the development of a strategy for the marine safety investigation in liaison with other substantially interested States;

.4 ensuring the methodology followed during the marine safety investigation is consistent with that recommended in resolution A.884(21);

.5 ensuring the marine safety investigation takes into account any recommendations or instruments published by the Organization or International Labour Organization, relevant to conducting a marine safety investigation; and

.6 ensuring the marine safety investigation takes into account the safety management procedures and the safety policy of the operator of a ship in terms of the ISM Code.

21.3 The marine safety investigating State(s) should allow a substantially interested State to participate in aspects of the marine safety investigation relevant to it, to the extent practicable.*

21.3.1 Participation should include allowing representatives of the substantially interested State to:

.1 interview witnesses;

.2 view and examine evidence and make copies of documents;

.3 make submissions in respect of the evidence, comment on and have their views properly reflected in the final report; and

.4 be provided with the draft and final reports relating to the marine safety investigation.

21.4 To the extent practicable, substantially interested States should assist the marine safety investigating State(s) with access to relevant information for the marine safety investigation. To the extent practicable, the investigator(s) carrying out a marine safety investigation should also be afforded access to Government surveyors, coastguard officers, ship traffic service operators, pilots and other marine personnel of a substantially interested State.

21.5 The flag State of a ship involved in a marine casualty or marine incident should help to facilitate the availability of the crew to the investigator(s) carrying out the marine safety investigation.

Chapter 22
Collection of evidence

22.1 A marine safety investigating State(s) should not unnecessarily detain a ship for the collection of evidence from it or have original documents or equipment removed unless this is essential for the purposes of the marine safety investigation. Investigators should make copies of documents where practicable.

22.2 Investigator(s) carrying out a marine safety investigation should secure records of interviews and other evidence collected during a marine safety investigation in a manner which prevents access by persons who do not require it for the purpose of the investigation.

---

* The reference to 'extent practicable' may be taken to mean, as an example, that co-operation or participation is limited because national laws make it impractical to fully co-operate or participate.

22.3 Investigator(s) carrying out the marine safety investigation should make effective use of all recorded data including voyage data recorders if fitted. Voyage data recorders should be made available for downloading by the investigator(s) carrying out a marine safety investigation or an appointed representative.

22.3.1 In the event that the marine safety investigating State(s) do not have adequate facilities to read a voyage data recorder, States with such a capability should offer their services having due regard to the:

.1 available resources;

.2 capabilities of the readout facility;

.3 timeliness of the readout; and

.4 location of the facility.

Chapter 23
Confidentiality of information

23.1 States should ensure that an investigator(s) carrying out a marine safety investigation only discloses information from a marine safety record where:

.1 it is necessary or desirable to do so for transport safety purposes and any impact on the future availability of safety information to a marine safety investigation is taken into account; or

.2 as otherwise permitted in accordance with this Code.*

23.2 States involved in a marine safety investigation under this Code should ensure that any marine safety record in their possession is not disclosed in criminal, civil, disciplinary or administrative proceedings unless:

.1 the appropriate authority for the administration of justice in the State determines that any adverse domestic or international impact that the disclosure of the information might have on any

---

* States recognize that there are merits in keeping information from a marine safety record confidential where it needs to be shared with people outside the marine safety investigation for the purpose of conducting the marine safety investigation. An example is where information from a marine safety record needs to be provided to an external expert for their analysis or second opinion. Confidentiality would seek to ensure that sensitive information is not inappropriately disclosed for purposes other than the marine safety investigation, at a time when it has not been determined how the information will assist in determining the contributing factors in a marine casualty or marine incident. Inappropriate disclosure may infer blame or liability on the parties involved in the marine casualty or marine incident.

current or future marine safety investigation is outweighed by the public interest in the administration of justice; and*

.2 where appropriate in the circumstances, the State which provided the marine safety record to the marine safety investigation authorizes its disclosure.

23.3 Marine safety records should be included in the final report, or its appendices, only when pertinent to the analysis of the marine casualty or marine incident. Parts of the record not pertinent, and not included in the final report, should not be disclosed.

23.4 States need only supply information from a marine safety record to a substantially interested State where doing so will not undermine the integrity and credibility of any marine safety investigation being conducted by the State or States providing the information.

23.4.1 The State supplying the information from a marine safety record may require that the State receiving the information undertake to keep it confidential.

Chapter 24
Protection for witnesses and involved parties

24.1 If a person is required by law to provide evidence that may incriminate them, for the purposes of a marine safety investigation, the evidence should, so far as national laws allow, be prevented from admission into evidence in civil or criminal proceedings against the individual.

24.2 A person from whom evidence is sought should be informed about the nature and basis of the investigation. A person from whom evidence is sought should be informed, and allowed access to legal advice, regarding:

.1 any potential risk that he/she may incriminate himself/herself in any proceedings subsequent to the marine safety investigation;

.2 any right not to self-incriminate or to remain silent;

---

* Examples of where it may be appropriate to disclose information from a marine safety record in criminal, civil, disciplinary or administrative proceedings may include:
  1 where a person, the subject of the proceedings, has engaged in conduct with the intention to cause a destructive result; or
  2 where a person, the subject of the proceedings, has been aware of a substantial risk that a destructive result will occur and, having regard to the circumstances known to him or her, it is unjustifiable to take the risk.

.3 any protections afforded to the person to prevent the evidence being used against him/her if he/she provides the evidence to the marine safety investigation.

Chapter 25
Draft and final report

25.1 Marine safety investigation reports from a marine safety investigation should be completed as quickly as practicable.

25.2 Where it is requested, and where practicable, the marine safety investigating State(s) should send a copy of a draft marine safety investigation report for comment to interested parties. However, this recommendation does not apply where there is no guarantee that the interested party will not circulate, nor cause to circulate, publish or give access to the draft marine safety investigation report, or any part thereof, without the express consent of the marine safety investigating State(s).

25.3 The marine safety investigating State(s) should allow the interested party 30 days or some other mutually agreed time to submit their comments on the marine safety investigation report. The marine safety investigating State(s) should consider the comments before preparing the final marine safety investigation report and where the acceptance or rejection of the comments will have direct impact on the interests of the interested party that submitted them, the marine safety investigating State(s) should notify the interested party of the manner in which the comments were addressed. If the marine safety investigating State(s) receives no comments after 30 days or the mutually agreed period has expired, then it may proceed to finalize the marine safety investigation report.*

25.4 Where it is permitted by the national laws of the State preparing the marine safety investigation report, the draft and final report should be prevented from being admissible in evidence in proceedings related to the marine casualty or marine incident that may lead to disciplinary measures, criminal conviction or the determination of civil liability.

25.5 At any stage during a marine safety investigation interim safety measures may be recommended.

---

* See chapter 13 where provisions with respect to providing interested parties with reports on request may alternatively be included as a mandatory provision.

25.6 Where a substantially interested State disagrees with the whole or a part of a final marine safety investigation report, it may submit its own report to the Organization.

Chapter 26
Re-opening an investigation

26.1 The marine safety investigating State(s) which has completed a marine safety investigation should reconsider its findings and consider re-opening the investigation when new evidence is presented which may materially alter the analysis and conclusions reached.

26.2 When significant new evidence relating to any marine casualty or marine incident is presented to the marine safety investigating State(s) that has completed a marine safety investigation, the evidence should be fully assessed and referred to other substantially interested States for appropriate input.

Resolution MSC.255(84)
(adopted on 16 May 2008)

Adoption of the Code of the International Standards and Recommended Practices for a Safety Investigation into a Marine Casualty or Marine Incident (Casualty Investigation Code)

THE MARITIME SAFETY COMMITTEE,

RECALLING Article 28(b) of the Convention on the International Maritime Organization concerning the function of the Committee,

NOTING with concern that, despite the best endeavours of the Organization, casualties and incidents resulting in loss of life, loss of ships and pollution of the marine environment continue to occur,

NOTING ALSO that the safety of seafarers and passengers and the protection of the marine environment can be enhanced by timely and accurate reports identifying the circumstances and causes of marine casualties and incidents,

NOTING FURTHER the importance of the United Nations Convention on the Law of the Sea, done at Montego Bay on 10 December 1982, and of the customary international law of the sea,

NOTING IN ADDITION the responsibilities of flag States under the provisions of the International Convention for the Safety of Life at Sea, 1974 (regulation I/21) (hereinafter referred to as "the Convention"), the International Convention on Load Lines, 1966 (article 23) and the International Convention for the Prevention of Pollution from Ships, 1973 (article 12), to conduct casualty investigations and to supply the Organization with relevant findings,

CONSIDERING the need to ensure that all very serious marine casualties are investigated,

CONSIDERING ALSO the Guidelines on fair treatment of seafarers in the event of a maritime accident (resolution A.987(24)),

ACKNOWLEDGING that the investigation and proper analysis of marine casualties and incidents can lead to greater awareness of casualty causation and result in remedial measures, including better training, for the purpose of enhancing safety of life at sea and protection of the marine environment,

RECOGNIZING the need for a code to provide, as far as national laws allow, a standard approach to marine casualty and incident investigation with the objective of preventing marine casualties and incidents in the future,

RECOGNIZING ALSO the international nature of shipping and the need for co-operation between Governments having a substantial interest in a marine casualty or incident for the purpose of determining the circumstances and causes thereof,

NOTING resolution MSC.257(84) by which it adopted amendments to chapter XI-1 of the Convention to make parts I and II of the Code of the International Standards and Recommended Practices for a Safety Investigation into a Marine Casualty or Marine Incident mandatory under the Convention,

HAVING CONSIDERED, at its eighty-fourth session, the text of the proposed Casualty Investigation Code,

1. ADOPTS the Code of the International Standards and Recommended Practices for a Safety Investigation into a Marine Casualty or Marine Incident (Casualty Investigation Code), set out in the annex* to the present resolution;

2. INVITES Contracting Governments to the Convention to note that the Code will take effect on 1 January 2010 upon entry into force of the amendments to regulation XI-1/6 of the Convention;

3. REQUESTS the Secretary-General of the Organization to transmit certified copies of the present resolution and the text of the Code contained in the annex to all Contracting Governments to the Convention;

4. FURTHER REQUESTS the Secretary-General of the Organization to transmit copies of the present resolution and the text of the Code contained in the annex to all Members of the Organization which are not Contracting Governments to the Convention.

---

* See page 1.

Notes

Notes